The Magic School Bus
Inside a Hurricane

The Magic School Bus
Inside a Hurricane

By Joanna Cole
Illustrated by Bruce Degen

André Deutsch Children's Books

The author and illustrator wish to thank Dr Robert C Sheets,
Director of the National Hurricane Centre; and Dr Daniel Leathers, Delaware State Climatologist,
University of Delaware, for their assistance in preparing this book.

Scholastic Children's Books,
Commonwealth House, 1-19 New Oxford Street,
London WC1A 1NU, UK
A division of Scholastic Ltd
London – New York – Toronto – Sydney – Auckland

First published in the US by Scholastic Inc. 1995
This edition published by Scholastic Ltd, 1996

ISBN: 0 590 54255 9

She's the Magic School Bus Editor
And we'd like herewith to credit her:
To Phoebe Yeh
Our Sunny Day.

J.C. & B.D.

We were learning about weather.
Absolutely everything in our room
was about rain or snow or sun or wind.
Every kid in the class
was doing a weather project.
We were even listening to
weather reports on Miss Frizzle's radio.

We started going up, and Miss Frizzle said, "Didn't I mention, children, that hot air rises?"

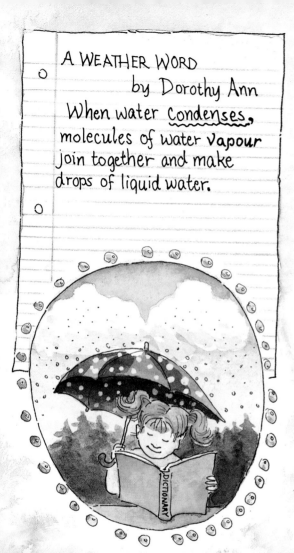

A WEATHER WORD
        by Dorothy Ann
When water <u>condenses</u>, molecules of water **vapour** join together and make drops of liquid water.

"Warm air rising from earth carries lots of water vapour molecules," Miss Frizzle continued.
"As the air rises, it cools down. The water condenses in the air and forms clouds."

DID YOU BRING YOUR RAINCOAT, ARNOLD?

TELL ME THIS ISN'T HAPPENING....

We drifted into the centre of a cloud.
Miss Frizzle was right – it was *damp* in there.
The cloud was made of tiny water droplets
hanging in the air.

Down below, the weather forecasters were standing in the rain.
They didn't see us inside the cloud, but we could hear their voices.
One of them said, "I hope that teacher knows there's a *hurricane watch* in effect."

CHECK OUT MY HURRICANE WATCH, ARNOLD. GET IT? HURRICANE WATCH!!

I'M PRETENDING I CAN'T HEAR....

WHAT IS A HURRICANE?
by Florrie
A hurricane is one of the most violent kinds of storms.
In a hurricane, winds swirl in a circle around the storm's centre at 74 miles per hour or more!

HURRICANE SYMBOL

MORE WORDS FROM DOROTHY ANN
A Hurricane Watch means that a hurricane may strike within the next 36 hours.
A Hurricane Warning means that a hurricane is likely to strike within the next 24 hours.
A warning is more urgent than a watch.

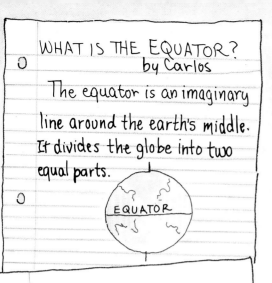

WHAT IS THE EQUATOR?
by Carlos

The equator is an imaginary line around the earth's middle. It divides the globe into two equal parts.

EQUATOR

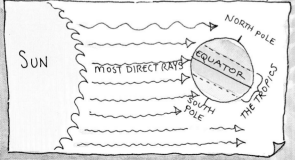

WHY IS IT HOTTER NEAR THE EQUATOR?
by Michael

Because of the way the earth is tilted, the sun's rays almost always shine towards the earth's middle. This means there are no cold winters there.

SUN
MOST DIRECT RAYS
NORTH POLE
EQUATOR
SOUTH POLE
THE TROPICS

As usual, Miss Frizzle paid no attention.
She turned up the fire, and more
hot air rushed into the balloon.
As we rose above the cloud,
the wind started pushing us south.
Before long, we had travelled thousands of miles.
Miss Frizzle said we were above a tropical ocean
about five hundred miles north of the equator.

WOW! LOOK AT THAT WATER!

WE CAN GO SWIMMING!

AND WINDSURFING!

AND SNORKELLING!

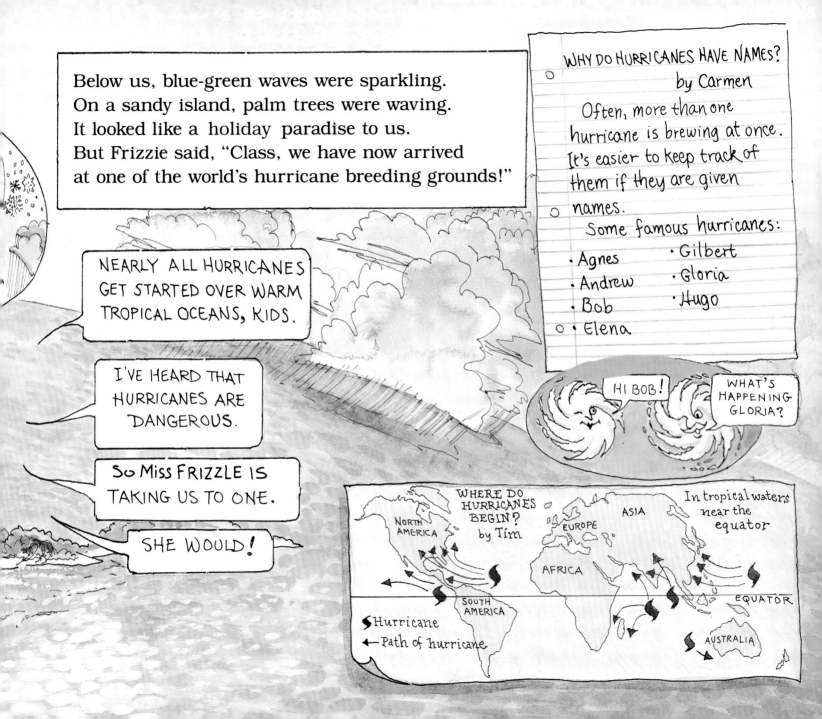

Below us, blue-green waves were sparkling.
On a sandy island, palm trees were waving.
It looked like a holiday paradise to us.
But Frizzie said, "Class, we have now arrived
at one of the world's hurricane breeding grounds!"

WHY DO HURRICANES HAVE NAMES?
by Carmen

Often, more than one hurricane is brewing at once. It's easier to keep track of them if they are given names.

Some famous hurricanes:
- Agnes
- Andrew
- Bob
- Elena
- Gilbert
- Gloria
- Hugo

NEARLY ALL HURRICANES GET STARTED OVER WARM TROPICAL OCEANS, KIDS.

I'VE HEARD THAT HURRICANES ARE DANGEROUS.

So Miss FRIZZLE IS TAKING US TO ONE.

SHE WOULD!

HI BOB!

WHAT'S HAPPENING GLORIA?

WHERE DO HURRICANES BEGIN?
by Tim

In tropical waters near the equator

NORTH AMERICA
EUROPE
ASIA
AFRICA
SOUTH AMERICA
EQUATOR
AUSTRALIA

$ Hurricane
← Path of hurricane

WHEN IS HURRICANE SEASON?
by Rachel

Most hurricanes begin in the late summer and early autumn. That is when tropical oceans are warmest.

The warmer the ocean is, the stronger the hurricane is likely to become.

"Class, remember that as hot air rises from the ocean surface, the water vapour in the air condenses and forms clouds," said the Friz.
Down below, more hot air rushed in from all sides to take the place of the rising air.
In the middle of the rising air, a column of sinking air formed.
We started sinking with it.

"Oh dear," said Miss Frizzle.
"The balloon must have sprung a leak."
Hot air was rushing out, and the balloon
was plummeting fast.
"Jump ship, class!" shouted the Friz.
She jumped overboard, and we went after her.
Right away, we knew it was a big mistake.

DO ALL TROPICAL STORMS
BECOME HURRICANES?
by Amanda Jane
No. All around the
world, there are more than
100 tropical storms each
year. Only about 60 of them
grow to hurricane strength.
And only a handful
of those ever reach places
where people live.

HURRY UP AND
JUMP, ARNOLD!

I CAN'T LOOK!

FOLLOW
ME, KIDS!

WHAT MAKES HURRICANE WINDS BLOW IN A CIRCLE?
by Alex

Winds begin by blowing straight. But the movement of the earth as it spins on its axis makes them curve.

WINDS CURVE

THE EARTH SPINS →

AXIS

The faster winds blow, the more they curve. Hurricane winds are very fast, so they curve and curve until they make a circle.

The wind was blowing the clouds into a huge circle.
"The storm is starting to take on the typical shape of a hurricane. Isn't it fascinating, children?" shouted Miss Frizzle.

It was more than fascinating.
It was terrifying!
We were caught in the edge of the storm,
blowing around and around in a giant whirlwind.
That whirlwind was a hurricane!

HOW BIG IS A HURRICANE?
by John
Hurricanes are enormous.
Each one is about 10 miles
high and 300 to 600 miles
wide!

A TYPICAL HURRICANE
HAS A LIFE SPAN
OF ABOUT 10 DAYS.

LISTENERS— WE'LL
BE TELLING YOU
ABOUT THE
WHOLE HURRICANE.

MAYBE ITS
BATTERIES WILL
RUN OUT SOON.

Where We Are in the Hurricane

All around were columns of air
called hot towers, or chimneys.
They were sucking up more and more
hot moist air from the ocean.
The heat energy from the air
was feeding the storm
and making it stronger.
The plane was shaking
and so were we!

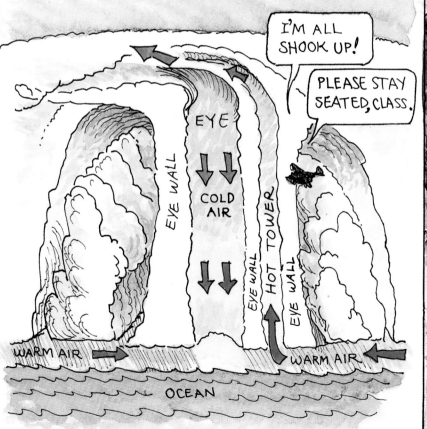

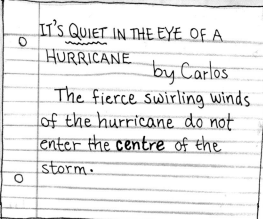

IT'S QUIET IN THE EYE OF A HURRICANE
                          by Carlos
    The fierce swirling winds
of the hurricane do not
enter the **centre** of the
storm.

WINDS   WINDS

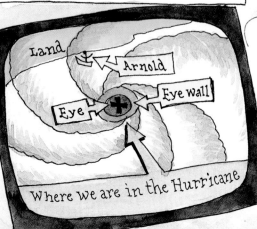

Land
Arnold
Eye   Eye wall
Where we are in the Hurricane

Then suddenly everything was quiet.
"Class, we have entered the eye – or centre of the hurricane!" announced Miss Frizzle.
The ocean waves still crashed below
and the winds howled outside,
but in the eye only gentle breezes blew.
Up above, the sky was blue
and the sun was shining.
We relaxed and enjoyed ourselves.

PEACE AND QUIET!

BALMY BREEZES!

A-A-AH!

We flew about thirty miles
across the eye.
Then the Friz called out,
"We will enter the other side
of the eye wall now."
"Don't go!" we cried,
but the plane was already
on its way — back into the
hurricane's fierce wind and rain.

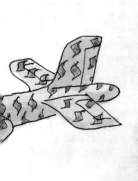

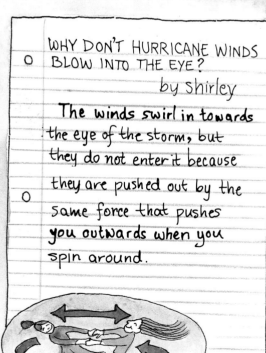

WHY DON'T HURRICANE WINDS
BLOW INTO THE EYE?
by shirley

The winds swirl in towards
the eye of the storm, but
they do not enter it because
they are pushed out by the
same force that pushes
you outwards when you
spin around.

HOW HURRICANES TRAVEL
by Wanda

When a hurricane starts, it usually moves slowly— about 10 to 20 miles per hour. As the storm gets **further** north, its speed can increase up to 60 miles per hour! Hurricanes can travel hundreds of miles **each day**.

SPEED 25 LIMIT

WHICH PART OF THE HURRICANE IS STRONGEST?
by Florrie

The right front corner is strongest because the whirling winds are circling **towards the shore.** They add their strength to the winds that move the storm forward.

The entire hurricane was moving across the ocean towards land, and we were going with it! "The right forward corner of the hurricane as you are looking towards land has the strongest wind and rain and the highest ocean waves," shouted the Friz. Naturally, she flew directly into that part.

A HURRICANE MOVES LIKE A TOP SPINNING ACROSS THE FLOOR.

IT MOVES TWO WAYS—

IT SPINS AROUND...

...AND, IT TRAVELS FORWARDS.

Most damage will be done here

LAND

STORM'S LEFT FRONT

EYE

STORM'S RIGHT FRONT

RIGHT REAR

LEFT REAR

FORWARD STORM MOVEMENT

HURRICANES...THEN AND NOW
by Ralphie

In the past, there was less property **damage**. Today, coastal areas are more built up. So more houses and buildings are damaged in hurricanes.

But today, not as many lives are lost! In past times, many people were killed because no one knew when hurricanes were coming. Today, weather forecasts tell us to get ready for hurricanes.

As the hurricane approached land,
the wind pulled up trees by the roots
and blew the roofs off houses.
It also blew ashore a huge dome of water
called the storm surge.
The ocean rose ten feet higher than usual
and, on top of that, there were giant waves.
We were horrified as we watched the storm surge
sweep over the shore below.

IN 1900, MORE THAN 6,000 PEOPLE DROWNED WHEN A STORM SURGE SWEPT ACROSS GALVESTON ISLAND, TEXAS, CLASS.

THAT WAS A LONG TIME AGO.

IT WOULDN'T HAPPEN NOW.

SANDBAGS

BOARD UP WINDOWS

EVACUATE

But that was nothing compared to the horror we felt when we heard the Friz shouting above the sound of roaring water, "We seem to be running out of fuel, children!" Sure enough, the plane was dipping lower and lower.

As we fell into the water, we saw Arnold waving to us from a nearby roof.

Somehow Arnold managed to get on the plane
before we were swept away by the waves
at the front edge of the hurricane.
The water was creeping up the windows.
The plane was going to sink for sure!
Then we saw a dark, funnel shape coming our way.

"I've seen that shape on TV," said Ralphie.
"I read about it in a book!" said Keesha.
The twister came right for us.
The next thing we knew, it had picked us up,
and we were travelling by tornado!

TORNADOES OFTEN OCCUR AT THE EDGES OF HURRICANES THAT ARE MOVING OVER LAND, CLASS.

ARE TORNADOES AND HURRICANES ALIKE?
by Phil

Yes and no.
Tornadoes and hurricanes are both whirlwinds.
But tornadoes:
1. are much smaller than hurricanes
2. have faster winds, for the most part
3. destroy almost everything in their path.

Tornadoes can twist at speeds of 200 to 300 miles per hour.

A TYPICAL TORNADO HAS A SHORT LIFE SPAN— ONLY A FEW MINUTES.

I THINK MY LIFE SPAN JUST GOT SHORTER.

CAN TORNADOES REALLY CARRY OBJECTS?
by Keesha

Yes! Tornadoes are like giant vacuum cleaners. They lift dirt, rubbish, and even large objects, such as houses, cars, trees, and **railway trains!**

Once a tornado picked up a crate of eggs and set them down miles away. Not one egg was broken!

After a while we felt a bump and looked around.
The tornado had set us down gently.
We were in our old school bus again.
We were dressed in our usual clothes again.
The hurricane was over.
And we were at a petrol station.

THANK GOODNESS!

USUALLY THE THINGS THAT ARE PICKED UP BY A TORNADO ARE BROKEN APART...

BUT NOT ALWAYS.

CLUCK CLUCK EGGS

No CRACKS No QUACKS

SUPER

MAGIC

REGULAR

SUPER

MAGIC

Miss Frizzle filled up the tank
and drove down the road
as if nothing had happened.
"As I said earlier, class, we are on our way
to visit a weather station," she said.

The weather forecasters at the station had a lot to tell us about hurricanes. *We* had a lot to tell *them*, too!

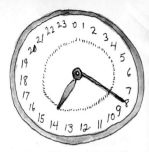

Finally, we drove back to school and finished our weather projects.

After that trip, we needed some time to relax.
Miss Frizzle said we could have a party.
We had great games, crazy dancing, and yummy snacks.
And for a while, we didn't even think about
Miss Frizzle's next class trip!

The Magic School Bus Mail Bag

Letters... we get letters...

To the Magic School Bus Editor:
You should not have said that a school bus could turn into a hot air balloon or a weather plane. That cannot really happen.
Your friend,
Sam

EXOTIC BROOKLYN
FOR SALE

Dear Joanna,
Radios cannot have conversations with people.
Barbara

To: JOANNA COLE
AUTHOR
c/o Scholastic Inc.

GREETINGS FROM SUNNY EAST ORANGE, N.J.

Dear Joanna and Bruce,
Reading about hurricanes may be fun, but it is no fun to be in one!
I know because my family was in Hurricane ANDREW and it was scary!
– Keith

Dear Bruce,
Radios do not dance.
from Jean

To: Bruce Degen
ARTIST
c/o Scholastic Inc.

Dear Arnold,
On your trip, the hurricane reached land. But most hurricanes go far out to sea and do not hurt people and property.
Your friend,
Al, the weather scientist

A fishing boat probably would not survive if it were out in a very strong hurricane.
From the Coast Guard

Dear Bruce,
If Arnold really fell from a great height into the ocean, he would need medical attention!
From your Doctor

Dear Joanna,
The things that happen on Miss Frizzle's trips are much too risky for children. Please keep them home next time. Your Mother

Winter in Connecticut

MSB CENTRAL

MOONLIGHT IN BURBANK

MSB

Miss Frizzle,
We think the whole class should enroll in Phoebe's old school.
—The students at the Better-Safe-Than-Sorry School

To all readers:
Some of the things that happen in this book are make-believe. But of course, all the Science is real!
Joanna & Bruce